John Gould

An Introduction to the Mammals of Australia

John Gould

An Introduction to the Mammals of Australia

ISBN/EAN: 9783337313784

Printed in Europe, USA, Canada, Australia, Japan

Cover: Foto ©berggeist007 / pixelio.de

More available books at **www.hansebooks.com**

INTRODUCTION

TO THE

MAMMALS OF AUSTRALIA.

BY

JOHN GOULD F.R.S.

ETC. ETC.

LONDON:

PRINTED FOR THE AUTHOR,

26 CHARLOTTE STREET, BEDFORD SQUARE,

BY TAYLOR AND FRANCIS, RED LION COURT, FLEET STREET.

1863.

TO

HIS ROYAL HIGHNESS

THE PRINCE CONSORT

THIS WORK

ON THE

MAMMALS OF AUSTRALIA

IS,

WITH HIS ROYAL HIGHNESS'S PERMISSION,

DEDICATED

BY HIS MOST OBLIGED AND

FAITHFUL SERVANT

JOHN GOULD.

Having been permitted to dedicate my work on the "Birds of Australia" to Her Most Gracious Majesty Queen Victoria, I was naturally desirous of dedicating the companion-work on the Mammals of the same country to Her Majesty's most enlightened and accomplished Consort; and the required permission was readily and graciously granted me. The dispensation which has deprived Her Majesty and the Prince's adopted country of one whose untimely loss we all deplore, still leaves me the privilege of that permission, and my work will continue to have the honour of being inscribed to His Royal Highness. It is with a melancholy satisfaction that I accordingly retain that Dedication, which, should it meet my Sovereign's eye, will, I think, only recall to her that love which the whole country entertains for his cherished memory. I feel that nothing I can say respecting the admirable qualities of this most enlightened Prince can in any way add to the deservedly high reputation of one whose great learning and manifold virtues, while he was among us, did so much for Science and Art, and whose example, we trust, will influence generations yet unborn.

NOTICE.

———————

The Preface and Introduction to my "Mammals of Australia" having been set up in small type for facility of correction, I have had a limited number of copies printed in an octavo form, for distribution among my scientific friends and others, to whom I trust it will be at once useful and acceptable. They must however still regard it rather in the light of a proof-sheet, inasmuch as it contains many imperfections, which have been corrected in the folio edition.

PREFACE.

In the Preface to the 'Birds of Australia,' which has now been fifteen years before the public, I stated that, "Having in the summer of 1837 brought my work on the 'Birds of Europe' to a successful termination, I was naturally desirous of turning my attention to the Ornithology of some other region; and a variety of opportune and concurring circumstances induced me to select that of Australia, the birds of which country, although invested with the highest degree of interest, had been almost entirely neglected." But if the Birds of Australia had not received that degree of attention from the scientific ornithologist which their interest demanded, I can assert, without fear of contradiction, that its highly curious and interesting Mammals had been still less investigated. It was not, however, until I arrived in the country, and found myself surrounded by objects as strange as if I had been transported to another planet, that I conceived the idea of devoting a portion of my attention to the mammalian class of its extraordinary fauna.

The native black, while conducting me through the forest or among the park-like trees of the open

b 2

plains, would often point out the pricking of an Opossum's nails on the bark of a *Eucalyptus* or other tree, and indicate by his actions that in yonder hole, high up, was sleeping an Opossum, a *Phalangista*, or a Flying *Petaurus*. Even the objects brought to our bush-fires were enough to incite a desire for a more extended knowledge of Australia's Mammals; for numerous were the species of Kangaroos and Opossums that were nightly roasted and eaten by these children of nature. Perchance a half-charred log, or the heated hollow branch of a *Eucalyptus*, would send forth into the lap of one or other of the surrounding guests the *Acrobates pygmæus*, the white-footed *Hapalotis*, or other small quadruped. Tired by a long and laborious day's walk under a burning sun, I frequently encamped for the night by the side of a river, a natural pond, or a water-hole, and before retiring to rest not unfrequently stretched my weary body on the river's bank; while thus reposing, the surface of the water was often disturbed by the little concentric circles formed by the *Ornitho-rhynchus*, or perhaps an *Echidna* came trotting up towards me. With such scenes as these continually around me, is it surprising that I should have entertained the idea of collecting examples of the indigenous Mammals of a country whose ornithological productions I had gone out expressly to investigate? To have attempted to acquire a knowledge of more than the Birds and Mammals would have been unwise; still I was not insensible to the interest which attaches to its insects and to its wonderful botanical productions. The *Eucalypti*, the *Banksiæ*, the *Casuarinæ*, the native Cedar- and the Fig-trees will ever stand forth prominently in my memory. While in the interior of the country, I formed the intention of publishing a monograph of the great family of Kangaroos; but soon after my return to England I determined to attempt a more extended work, under the title of the 'Mammals of Australia.'

It will always be a source of pleasure to me to remember that I was the first to describe and figure the Great Black and Red Wallaroos (*Osphranter robustus* and *O. antilopinus*), the three species of *Onychogalea*, several of the equally singular *Lagorchestes,* and many other new species of Kangaroos. Mounted examples of all these animals, whether discovered by myself or by others, are now contained in the national collection of this country; but I regret to say that their colours are very different from what they were while the animals were living, the continuous exposure to light, consequent upon their being placed in a museum, causing their evanescent colouring rapidly to fade, both here and in the collections of every other country. Those who have seen the living *Osphranter rufus* at the Zoological Gardens could scarcely for a moment suppose that the Museum specimen of the same animal had ever been dressed in such glowing tints. To see the Kangaroos in all their glory, their native country must be visited; their beauty would then be at once apparent, and their various specific distinctions easily recognizable.

The exploration of every new district has afforded ample proof of the existence of species in every department of zoology with which we were previously unacquainted. Under these circumstances, I do not consider my work to be in any way complete, or that it comprises nearly the whole of the Mammals of a country of which so much has yet to be traversed; but I bring it to a close after an interval of eighteen years since its commencement, during which constant attention has been given to the subject, as treating upon the genera and species known up to the present time. If my life be prolonged, and the blessing of health be continued to me, I propose, as in the case of the ' Birds of Australia,' to keep the subject complete, by issuing a supplementary part, from time to time, should sufficient new materials be acquired to enable me so to do.

As with regard to my other publications, so also with this, I have to offer my best thanks to many persons for the kind and friendly assistance they have rendered me in prosecuting my labours on the ' Mammals of Australia.' I cannot, therefore, close these remarks without recording my obligations to Professor Owen, Dr. Gray, and G. R. Waterhouse, Esq., of the British Museum; to Ronald C. Gunn, Esq., of Launceston; the Rev. T. J. Ewing and Dr. Milligan of Hobart Town; to Dr. Bennett, W. S. MacLeay, Esq., Gerard Krefft, Esq., the late Dr. Ludwig Becker, W. S. Wall, Esq., the authorities of the Australian Museum, and the late Frederick Strange, of New South Wales; to Charles Coxen, Esq., of Queensland; John

Macgillivray, Esq.; the late Commander J. M. R. Ince, R.N.; to His Excellency Sir George Grey, formerly Governor of South Australia, and now of New Zealand; the late John Gilbert; Professor M'Coy, of Melbourne; George French Angas, of Angaston, South Australia; W. Ogilby, Esq., formerly Secretary of the Zoological Society of London; Dr. Sclater, its present Secretary; R. F. Tomes, Esq.; M. Jules Verreaux, of Paris; Dr. W. Peters, of the Royal Museum of Berlin; and lastly, my son, Mr. Charles Gould, the Geological Surveyor of Tasmania. I believe I have here enumerated the names of all who have favoured me with specimens or with the benefit of their opinions, in reference to the subjects of the present work. To have omitted the name of one friend would be a source of much vexation to me; but if such should unfortunately have been done, I trust it will be considered the result of inadvertence, and not of intentional neglect.

To my artist, Mr. Richter, I consider (and I have no doubt my readers will concur in my opinion) that much credit is due for the manner in which he has executed the drawings, both from the dead as well as from the living examples from which they were taken. Of my secretary, Mr. Prince, I have also to speak as having discharged the same praiseworthy services as heretofore.

It will be observed that, in mentioning the localities frequented by the various species, I have mostly used the term Van Diemen's Land for the large island lying off the south coast of Australia; there is now, however, a very general desire that it should be called Tasmania—in honour of Tasman, its original discoverer; this term has, therefore, also been used, and hence has arisen the discrepancy of using two names for one island. Even since the commencement of the work, new colonies have sprung up, or the older ones have been divided; thus the country now known as Queensland was formerly part of New South Wales, and Victoria was, until lately, known as Port Phillip.

INTRODUCTION.

In the foregoing Preface I have glanced at the principal groups
of Mammals inhabiting the great country of Australia. It will
now, however, be necessary to enter into greater detail respecting
this division of its fauna; and I conceive that it will not be out
of place if I commence with a retrospective view of the gradual
discovery of countries and their zoological productions from the
earliest historic times. Such a retrospect will not, I think, be
deemed unnecessary, especially since my intention is to show to
the general reader, rather than to the scientific naturalist, that
each great division of the globe has its own peculiar forms of
animal life, and that the fauna of Australia is widely different
from that of every other part of the world. By a mere glance
at the zoological features of the globe as at present existing, it
will be perceived with what precision the animal life of each
country has been adapted to its physical character; the absence
of certain great families of birds and quadrupeds in some
countries will also be apparent. To account for this on any
scientific principle would be very difficult, when we cannot say
why the Nightingale is not a summer visitant to Devonshire,
or why the Grouse is not found south of Wales; why the aërial
Swifts, Swallows, and Martins are numerous in Australia, and
absent in New Zealand; or why Woodpeckers, which occur in
nearly every other part of the globe, are not found in Australia,
New Guinea, or any of the Polynesian Islands.

B

The ancient Egyptians appear to have been little acquainted with the natural productions of any other country than their own,—at least, we have no evidence that they were; for neither so conspicuous a bird as the Peacock, nor even the Common Fowl, are represented on their lasting monuments. Of the eastern countries Alexander's expedition doubtless greatly increased the knowledge of the Greeks, furnishing materials for the philosophic mind of Aristotle, and certainly extending the knowledge of Pliny, as is evidenced by his 'Historia Naturalis,' the only work which has come down to us of the latter great naturalist. Pliny, standing out as a bright star in zoological science at the period he lived, was doubtless tolerably acquainted with the natural productions of Eastern Europe, Arabia, North-eastern Africa, slightly with those of Persia, and still less so with those of India.

It may be fairly said, that the earliest dawn of natural history commenced with the Christian era,—Aristotle living just before, and Pliny soon after, the advent of our Saviour. This early dawn, however, was for a long period obscured by the dark ages which succeeded; for it was not until the commencement of the 17th century that Aldrovandus, Piso, Maregrave, and Willughby wrote their works on this branch of science. At this comparatively late period, the productions of Europe were better known; Africa had been for a long time circumnavigated, and its southern fauna partially brought to light; India also in like manner furnished her quota, though sparingly, to the stock of human knowledge. What Alexander's celebrated expedition did for the naturalists Aristotle and Pliny, the discoveries of Columbus did by shedding a new light upon zoological science, and furnishing fresh food to the modern writers above mentioned. Linnæus, the greatest of all systematists, had a very extended knowledge of the natural productions of the globe, and the information this great man has left behind him in his numerous writings is considerable. Still, the southern land which we designate Australia (the mammalian products of which

this work is intended to illustrate) was a sealed book to him.
As regards this great country, it may be said that its most
highly organized animals, if we except the Seals, are the various
species of Rodents, and the equally numerous insectivorous and
frugivorous Bats, both of which rank among the lowest of the
Placentals. In America the *Marsupialia* are but feebly represented; in Africa and India none of this form exist. On the
other hand, Australia is the great country of these pouched
animals; they are universally distributed throughout its entire
extent, from north to south, and from east to west; and they
are not even absent from the neighbouring islands. Their
presence in Tasmania on the south, and New Guinea on the
north, testifies that these countries were formerly united to the
mainland, and constituted a great natural province, characterized by a similar fauna and flora. It will be unnecessary for
me to state that none of the *Quadrumana*, or Monkeys, are found
in Australia; and that neither the Lion, the Tiger, the Leopard,
nor any other of the *Felinæ*, roam among its forests, to disturb
the harmony of its generally peaceful quadrupeds.

The great groups of the *Bovinæ*, or Oxen, the *Equinæ*, or
Horses and Zebras, the stately Elephant, the huge Rhinoceros,
as well as the *Cervidæ*, or Deer-kind, and the Antelopes, are
totally unknown in Australia; yet the great grassy plains and
other physical features of the country would appear to be well
adapted for them and also for the smaller herbivorous quadrupeds,
such as the Hare, the Rabbit, &c. Why there should occur so
great a difference between the animals of Australia and those of
the other countries of the world it is not for me to say. But I
may ask, has creation been arrested in this strange land? and, if
not, why are these higher types denied to it? Whatever opinion
may be formed on this interesting subject, it is generally believed
that no more highly organized animals than those which are
now found there ever roamed over her plains or tenanted her
luxuriant brushes. At the same time, the partially fossilized
remains of distinct species of Kangaroos which have been disco-

vered in her stalactitic caves, and the huge skeletons, or parts
of skeletons, which have been exhumed from her alluvial beds,
testify that Australia must be of remote origin.   It is scarcely
necessary to remark that all these remains belong to Marsupial
animals; nor must it be imagined that I am oblivious of the fact
that the remains of members of this group have been found in
the older tertiary and secondary strata of Europe.   I merely
glance at these things, and leave their consideration to those
who pay special attention to the sister science of geology.

    Although the more highly organized animals do not inhabit,
and seem never to have inhabited Australia, it is not a little
interesting to observe how completely the law of representation
is manifested among her mammals—how one family typifies
another in the higher groups of the *Placentalia*; or, to be more
explicit, to note how the *Herbivora* are represented by the
Kangaroos, the *Felinæ* by the *Dasyures*, the Jerboas by the
Hapalotides, &c.   When speaking of the wonderful fossil *Diprotodon*, in his work on Palæontology, Professor Owen states—
"Australia yields evidence of an analogous correspondence
between its last extinct and its present aboriginal mammalian
fauna, which is the more interesting on account of the very
peculiar organization of most of the native quadrupeds of that
division of the globe.   That the *Marsupialia* form one great
natural group is now generally admitted by zoologists; the
representatives in that group of many of the orders of the more
exclusive Placental subclass of the Mammalia of the larger continents have also been recognized in the existing genera and
species : the *Dasyures*, for example, play the parts of the *Carnivora*; the Bandicoots (*Perameles*), of the *Insectivora*; the Phalangers, of the *Quadrumana*; the Wombat, of the *Rodentia*; and
the Kangaroos, in a remoter degree, of the *Ruminantia*.   The
first collection of mammalian fossils from the ossiferous caves of
Australia brought to light the former existence on that continent of larger species of the same peculiar marsupial genera :
some, as the Thylacine, and the Dasyurine subgenus represented

by the *D. ursinus*, are now extinct on the Australian continent; but one species of each still exists on the adjacent island of Tasmania; the rest were extinct Wombats, Phalangers, Potoroos, and Kangaroos—some of the latter (*Macropus Atlas, M. Titan*) being of great stature. A single tooth, in the same collection of fossils, gave the first indication of the former existence of a type of the Marsupial group, which represented the Pachyderms of the larger continents, and which seems now to have disappeared from the face of the Australian earth,—of the great quadruped, so indicated under the name of *Diprotodon* in 1838; and successive subsequent acquisitions have established the true marsupial character and the near affinities of the genus to the Kangaroo (*Macropus*), but with an osculant relationship with the herbivorous Wombat. The entire skull of the *Diprotodon*, lately acquired by the British Museum, shows *in situ* the tooth on which the genus was founded. This skull measures 3 feet in length, and exemplifies by its size the huge dimensions of the primeval Kangaroo. Like the contemporary gigantic Sloth in South America, the *Diprotodon* of Australia, while retaining the dental formula of its living homologue, shows great and remarkable modifications of its limbs. The hind pair were much shortened and strengthened compared with those of the Kangaroo; the fore pair were lengthened, as well as strengthened. Yet, as in the case of the *Megatherium*, the ulna and radius were maintained free, and so articulated as to give the fore paw the rotatory actions. These, in *Diprotodon*, would be needed, as in the herbivorous Kangaroo, by the economy of the marsupial pouch. The dental formula of *Diprotodon* was the same as in *Macropus major*: the first of the grinding series was soon shed, but the other four two-ridged teeth were longer retained; and the front upper incisor was very large and scalpriform, as in the Wombat. The zygomatic arch sent down a process for augmenting the origin of the masseter muscle, as in the Kangaroo. The foregoing skull, with parts of the skeleton of the *Diprotodon australis*, were discovered in a lacustrine deposit, probably

pleistocene, intersected by creeks, in the plains of Darling
Downs, Australia.

"The same formation has yielded evidence of a somewhat
smaller extinct herbivorous genus (*Nototherium*), combining, with
essential affinities to *Macropus*, some of the characters of the
Koala (*Phascolarctos*).   The writer has recently communicated
descriptions and figures of the entire skull of the *Nototherium
Mitchelli* to the Geological Society of London.   The genus *Phascolomys* was at the same period represented by a Wombat (*P.
gigas*) of the magnitude of a Tapir.   The pleistocene marsupial
Carnivora presented the usual relations of size and power to the
Herbivora whose undue increase they had to check."

In another work, Prof. Owen represents an almost entire
skull, with part of the lower jaw, of an animal (*Thylacoleo*)
rivalling the Lion in size, the marsupial character of which is
demonstrated by the position of the lacrymal foramen in front
of the orbit, by the palatal vacuity, by the loose tympanic bone,
by the development of the tympanic bulla in the alisphenoid, by
the very small relative size of the brain, and other characters.
"The carnassial tooth is 2 inches 3 lines in longitudinal
extent, or nearly double the size of that in the Lion.   The
upper tubercular tooth resembles, in its smallness and position,
that in the placental Felines.   But in the lower jaw the carnassial is succeeded by two very small tubercular teeth, as in
Plagiaulax; and there is a socket close to the symphysis of the
lower jaw of *Thylacoleo*, which indicates that the canine may
have terminated the dental series there, and have afforded an
additional feature of resemblance to the *Plagiaulax*."

As might naturally be expected, the climate of a country
which extends over more than 30 degrees of latitude is very much
diversified.   Cape York and Arnheim's Land are as near 11°
south as possible, while Wilson's Promontory, in Victoria, reaches
39°, and the southern part of Tasmania $44\frac{1}{2}$°.   The parts of
Australia approaching the Tropic differ very considerably from
its southern portions; for, lying more to the north, the latter are

under the influence of monsoons, and rains more or less regular occur in their proper seasons. Speaking generally, however, Australia may be characterized as one of the driest and most heated countries of our globe; for, although an island in the strictest sense of the word, it is so extensive that the surrounding seas have little influence upon the distant interior, which must still be regarded as a great sterile waste, destitute of mountains sufficient to attract the moisture requisite to form navigable or other rivers. In writing this in 1863, when travellers have crossed the country and so many valuable discoveries have lately been made, I am willing to admit that this great desert is here and there relieved by higher lands which will ultimately become useful to the enterprising settler, and that, in all probability, many fine and extensive oases have yet to be brought to light; but, at the same time, I believe there will always be considerable uncertainty in the seasons of the interior of this great land. In southern latitudes we know that this is the case, while in the north a wet or a dry monsoon greatly alters the face of the country, and exerts a powerful influence on animal and vegetable life. Hence it is that the scanty fauna of this part of Australia is so organized that it is able to exist without water: the various species of Rodents, such as the members of the genera *Mus* and *Hapalotis*, and the Wombats, Lagorchestes, and Bettongias, and other Kangaroos, are thus constituted; and it will be recollected that, when speaking of the Halcyons and other large Kingfishers in the 'Birds of Australia,' I stated that I believed they never partook of this element, their food consisting of lizards and insects, to which, in like manner, it was not essential. The Australian mammals must, however, be put to severe straits occasionally, not from the want, but from the superabundance of water,—a wet monsoon in the north, and the heavy rains which occasionally occur in the south, deluging the basin-like surface of the interior and rendering it untenable, and obliging them to retire to the higher ridges until the drought, which generally ensues, has restored it to its normal condition. The districts,

or countries as I may call them, which constitute the other portions of Australia are very different, indeed completely opposite in
character; I mean the rich lands which surround nearly the whole
of the sterile centre. The mountain-ranges, of no very great
elevation it is true, exert much influence upon the face of nature,
constantly attracting rains, which, pouring down their sides, deposit a rich alluvial soil, favourable to the growth of gigantic trees
and the most luxuriant vegetation. The forests of Palms which
there occur are scarcely inferior to those of any other country,
while the stately native Cedars and Fig-trees are wonders to
every traveller. These giants of the forest are scarcely ever to be
found in the interior; sterility is not suited to their existence;
they do not occur in company with the *Banksiæ*, the *Hakeæ*, or
the *Casuarinæ*, most of which are characteristics of land wherein
the settler would not choose to risk his fortune. The great physical features of Australia then, as a whole, are the absence of
high mountains and navigable rivers, its heated interior, its
vast grassy plains, and its luxuriant brushes, particularly on its
southern and south-eastern coasts. Over the whole of this extensive country, with its ever-varying climate, certain groups
of animals are universally spread, while others, particularly the
more isolated forms, are strictly confined to their own districts,
each adapted for some special end and purpose,—as much as the
long bill of the Humming-bird (*Docimastes ensiferus*) is evidently
formed for exploring the lengthened tubular corollas of the
Brugmansiæ, or the greatly curved bill of two species of the same
family of birds (the *Eutoxeres Aquila* and *E. Condaminei*) is for
insertion into the honey-cups of the *Coryanthes speciosa* and its
allies,—or, to take a more striking instance, as the brush-like
tongues of the numerous honey-feeding Parrakeets and Honey-
eaters of Australia are constituted for obtaining the nectar from
the flowers of the universally spread and equally numerous
Eucalypti which form so prominent a feature in the flora of that
country.

    I will now give, as far as my knowledge of the subject will

permit, an enumeration of Australian mammals, the extent of their range, &c. In doing this, I shall commence with the Monotrematous section of the *Marsupiata*, which includes the *Ornithorhynchus* and two species of *Echidna*; I shall then proceed to the genera *Myrmecobius, Tarsipes, Chæropus, Peragalea, Perameles, Phascolarctos, Phalangista, Cuscus, Petaurista, Belideus, Phascogale, Sarcophilus, Dasyurus, Thylacinus,* and *Phascolomys*; and these will be followed by the great family of Kangaroos, with remarks upon their structural differences and the especial object for which these appear to have been designed; next we shall come to the feebly represented Placentals, the Seals, and Rodents; and, lastly, to the species of *Pteropus* and other Bats.

I have considered that, in a large illustrated work like the 'Mammals of Australia,' it would be out of place to enter into the anatomy of the objects I have represented. I have therefore omitted all details of this kind; neither have I included therein a repetition of the generic characters and Latin descriptions which have appeared in general works on Mammalogy, where they may be easily referred to. Those who wish to enter more fully into the generic characters of the Australian mammals will find all the information they can wish for in Mr. Waterhouse's valuable work, entitled 'A Natural History of the Mammalia,' a publication of such great promise and merit, that it becomes a matter of surprise and regret to all interested in this branch of science that the publisher decided upon not continuing it to its completion.

It will be observed that I have entirely omitted the Whales, Porpesses, and Dugong, my reason for so doing being that I had not sufficient opportunities for studying those animals in a state of nature, and have not therefore attempted that which I did not understand, and consequently could not have accomplished in a satisfactory manner. With regard to the Dugong, I must not omit thanking my relative, Charles Coxen, Esq., of Queensland, for his attention in sending me a skin and part of the skeleton of this animal; but even with these materials I found I could not produce an accurate representation of it in the living state. Although I do not inflict upon my readers the characters and distinctions of genera, I must not pass over unnoticed the principal features which distinguish the *Marsupiata* from the Placental Mammalia. In the first place, the former are considered to be much less highly organized than the latter: according to Professor Owen, the brain is deficient in both the corpus callosum and the septum lucidum; the cerebrum is small in proportion to the animal, contracted in front, and its surface is smooth, or presents but few convolutions; the cerebellum is entirely exposed, and has a vermiform process large in proportion to the lateral lobes; the olfactory lobes are large. Two venæ cavæ enter the heart; "the right auricle has no trace of a fossa ovalis." In point of fact, the main characteristic of the Marsupials, as distinguished from the Placentals, is that much of the embryotic life in the former is carried on in what may be called a sort of external uterus.

On my return from Australia, the venerable Geoffroy St.-Hilaire put the following question to me, "Does the Ornithorhynchus lay eggs?" and when I answered in the negative, that fine old gentleman and eminent naturalist appeared somewhat disconcerted. Now, this oviparous notion was nearly in accordance with the true state of things—somewhat akin to what is actually the case; and I consider the most striking peculiarity of this singular animal, and indeed of all the *Marsupiata*, to be the imperfectly formed state in which their young are born. The Kangaroo at its birth is not larger than a baby's little finger, which it is not very unlike in shape: in this extremely helpless state, the mother, by some means at present unknown, places this vermiform object to one of the nipples within her pouch or marsupium; by some equally unknown process, the little creature becomes attached by its imperfectly formed mouth to the nipple, and there remains dangling for days, and even weeks, during which it gradually assumes the likeness and structure of its parents; at length it drops from this lacteal attachment into the pouch, re-attaches itself when hunger prompts it so to do, and as often again tumbles off when its wants have been supplied. It is scarcely necessary to say that, after gaining sufficient strength, it leaves this natural pocket of the mother, leaps into the open air and sports about the plains or the forest, as the case may be, and returns again to its warm home, until at length the wearied mother denies it this indulgence and proceeds again to comply with the law which governs all creatures, that of reproduction. This is a very low form of animal life, indeed the lowest among the Mammalia, and exhibits the first stage beyond the development of the bird.

This description has reference not only to the Kangaroos, which mostly have but one young at a time, but is equally descriptive of the other members of this group, some of which have two, while others have three or four, and others, the *Phascogalæ* for instance, eight or nine at a birth; but in all cases, even with these large numbers, the young hang to the mammæ in the way I have described.

Independently of the low structure of the brain and the low form of reproduction of the Kangaroos, I ought to mention that two little bones have been expressly provided for the support of the marsupium; there is also a considerable difference in the dentition, as well as in the form of the lower jaw, by which this group of animals may at all times be distinguished. I have not failed to observe much disparity in size in the *Marsupiata*; they seem to be always growing; for the males get larger and still larger for years, even long after they have commenced the duty of reproduction, and hence individuals of all sizes occur, and occasionally one extraordinarily large may be met with. I have observed this in all the Marsupials, but particularly among the Kangaroos. The great herds of the grey species, *Macropus major*, are frequently headed by an enormous male, or Boomer as he is called. Like the "rogue Elephants" of Ceylon, these patriarchs are often solitary, and are generally very savage.

Commencing with the most lowly organized of the Australian mammals, I may state that the *Ornithorhynchus* has a very limited range, as is shown by its not being found either in Western or Northern Australia—the south-eastern portions of the continent and Van Diemen's Land being the localities to which it is confined.

The spiny *Echidna hystrix* has not yet been found to the northward of Moreton Bay on the east coast, and, except in New South Wales and the islands in Bass's Straits, it is very rare— so rare indeed, that I have never seen a specimen from South Australia; yet in all probability it will be found there, since Mr. Gilbert obtained an example at Swan River; this individual, however, did not come under my notice, and I am therefore unable to say if it were a true *E. hystrix*, or a western representative of that species.

The more hairy *Echidna setosa* is confined to Van Diemen's Land; but it is questionable whether it be really distinct from *E. hystrix*; the more southern position and colder climate of that island may have had the effect of giving it a warmer coat, whiter spines, and of altering its general appearance.

The single species representing the genus *Myrmecobius* (*M. fasciatus*) appears to be more plentiful in the Swan River Settlement than elsewhere; it nevertheless occurs in the Murray Scrub and other parts of South Australia, and from thence to the western coast it probably inhabits every locality suited to its habits and mode of life.

Like the *Myrmecobius*, the little honey-lapping *Tarsipes rostratus* stands quite alone—and a truly singular creature it is : to give the area over which it ranges is impossible, as we know far too little of these diminutive mammals to come to any positive conclusion on this point; at present, the neighbourhood of King George's Sound is one of the localities in which it has been seen in a state of nature.

Isolated in form and differing in the structure of its feet from every other known quadruped is the *Chæropus*, an animal which

frequents the hard grounds of the interior, over which it is
dispersed from New South Wales to Western Australia. The
specific term of *ecaudatus*, first applied to this animal in consequence of the specimen characterized being destitute of the
caudal appendage, must now sink into a synonym, that organ
being as well developed in this as in any other of the smaller
quadrupeds, the *Perameles* for instance, to which this singular
animal is somewhat allied.

The root-feeding Dalgyte, or *Peragalea lagotis*, leads us still
nearer to the genus *Perameles* : the fauna of Western Australia
is greatly enriched by the addition of this beautiful species.
I believe that South Australia may also lay claim to it; for
I have seen a tail, said to have been obtained *on the south
coast, which greatly resembled that of the Swan River *Peragalea*; but it may have pertained to an allied animal with which
we are not yet acquainted.

The members of the restricted genus *Perameles* are numerous
in species, and universally dispersed over the whole of Australia
and Van Diemen's Land; they also extend in a northerly direction to New Guinea and the adjacent islands. Of this genus
there are two well-marked divisions : one distinguished by bands
on their backs or crescentic markings across their rumps and by
their diminutive tails, the other by a uniformity in their colouring. The species of the former division inhabit the hot stony
ridges bordering the open plains; those of the latter the more
humid forests, among grass and other dense vegetation. Figures
of most of these Bandicoots, as they are called, and an account
of the manners, habits, and economy of each, so far as known,
will be found in their proper places in the body of the work.

The Phascogales, of which there are three, namely *P. penicillata*, *P. calura*, and *P. lanigera*, are all natives of the southern
portions of Australia, from east to west; they are, however,
rather denizens of the interior than of the provinces near the
coast, but the *P. penicillata* is alike found in both. Their dentition indicates that they are sanguinary in their disposition,—

a character which is confirmed by the *P. penicillata*, small as it comparatively is, being charged with killing fowls and other birds.

It might be thought that the *Phascogalæ* would naturally lead to the *Antechini*, but there is no real affinity between the two groups. I find it most difficult to arrange the Australian mammals in anything like a serial order; but the numerous species forming the genera *Antechinus* and *Podabrus* are, perhaps, as well placed here as elsewhere. Like the *Peramelides*, the members of those genera inhabit every part of Australia and the adjacent islands : the thick-tailed species, forming the genus *Podabrus*, frequent the interior rather than the coast; the *Antechini*, on the other hand, inhabit both districts; and wherever there are trees and shrubs, one or other of them may be found; some evince a partiality for the fallen boles lying on the ground, while others run over the branches of those that are still standing.

I now approach a better-defined section of the Australian Marsupiata than any of the preceding—the nocturnal Phalangers. These are divided into several genera—*Phascolarctos, Petaurista, Belideus, Phalangista, Cuscus, Acrobates,* and *Dromicia.* The extraordinary Koala is only found in the brushes of New South Wales. It stands quite alone—the solitary species of its genus, and it is well worth while to turn to my figures and description of this anomalous Sloth among the Marsupials. The *Petauristæ* are strictly brush-loving animals, and are almost entirely confined to New South Wales; some one or other of the *Belidei*, on the other hand, is found in all other parts of the Australian continent (except perhaps its western portion), wherever there are *Eucalypti* of sufficient magnitude for their branches to become hollow spouts wherein these nocturnes may sleep during the day. This form also occurs among the animals of the New Guinea group of islands. The little Opossum Mouse, *Acrobates pygmæus*, is a general favourite with the colonists; and well it may be so, for in its disposition it is as amiable as its form is elegant and its fur soft and beautiful: what the Dormouse is to the English boy, this little animal is to the juveniles of Australia. I have seen it kept as a pet, and its usual retreat in the day, while it sleeps, was a pill-box; as night approaches it becomes active, and then displays much elegance in its motions. The true *Phalangistæ* comprise many species; and are found in every colony, in Port Essington on the north, Swan River on the west, New South Wales and Queensland on the east, and Victoria and Van Diemen's Land on the south. They lead to the genus *Cuscus*, a form better represented in New Guinea and its islands than in Australia, where only one species has been discovered, in the neighbourhood of Cape York. Of the two fairy-like *Dromiciæ*, which live upon the stamens of flowers and the nectar of their corollas, one is found in Van Diemen's Land, the other in Western Australia. The description of a third species of this form has just been transmitted by Mr. Krefft to the Zoological Society; he states that it was taken from an example discovered by himself in New South Wales, and proposes to call it *D. unicolor.*

An equally remarkable and distinct division or group is composed of the Dasyures, to which the extraordinary *Sarcophilus ursinus* of Van Diemen's Land bears precisely the same degree of relationship that the Koala does to the Phalangers. Like the *Thylacinus*, the *Sarcophilus* is confined to Van Diemen's Land. And I would ask, why are these strange and comparatively large animals now restricted to so limited an area? for it can scarcely be supposed that they have not, at some time or other, inhabited the continent of Australia also. Had not Tasmania as well as the mainland been peopled for a long time by the human race, it might have been supposed that their extirpation from the continent had been effected by these children of nature. Whatever the cause may have been, it cannot now be ascertained, and we must be content to treat of the creatures that still exist. Of the true Dasyures, four very distinct species are dispersed over Australia from Van Diemen's Land to the shores of Torres' Straits. Tasmania is frequented by two (*Dasyurus maculatus* and *D. viverrinus*), the southern parts of the mainland by the same two species with the addition of a third (*D. Geoffroyi*), while the *D. hallucatus* inhabits the north. The animals of this genus are very viverrine both in their appearance and in their sanguinary disposition, and are probably the true representatives in Australia of that group of quadrupeds. The term ' sanguinary' is rightly applied to some of these animals, yet there is not one which a child might not conquer. The boldest of them are more troublesome than dangerous, and a robbery of the hen-roost is the utmost of the depredations their nature prompts them to commit.

I now come to the most bloodthirsty of the Australian mammals—the Wolf of the Marsupials—the *Thylacinus* of Tasmania's forest-clad country—the only member of its Order which gives trouble to the shepherd or uneasiness to the stockholder. Van Diemen's Land is the true and only home of this somewhat formidable beast, which occasionally deals out destruction among the flocks of the settler, to which it evinces a decided preference over the Brush Kangaroos, its more ancient food. To man, however, it is not an object of alarm; for the shepherd, aided by his dog, and stick in hand, does not for a moment hesitate about attacking and killing it. The large life-sized head and the reduced figures given in the body of the work well represent the *Thylacinus*, and all that is known of its habits will be found in the accompanying letter-press.

Until lately, only one species of *Phascolomys* or Wombat was clearly defined; but we now know that there are three, if not four, very distinct kinds; and in all probability others may yet be discovered, and prove that this form has a much more extended range than is at present supposed. The *P. Wombat* is still abundant in Van Diemen's Land and on some of the islands in Bass's Straits; and two or three species burrow in the plains of the southern countries of Australia generally. These huge, heavy, and short-legged animals, revelling in a state of obesity, feed most harmlessly on roots and other vegetable substances; they are the Rodents of their own Order, and the representatives of the Capybaras of South America. With this group I terminate the first volume; the next is devoted to the great family of the *Macropodidæ* or Kangaroos. This, the most important of all the Marsupial groups, both as to diversity of form and the number of species, is so widely and so universally dispersed over the Australian continent and its islands, that its members may be said to exist in every part of those countries. They are found in great abundance in the southern and comparatively cold island of Tasmania, while three species, at least, tenant that little-explored country, New Guinea, and some of the adjacent islands. Varied as the physical condition of Australia really is, forms of Kangaroos are there to be found peculiarly adapted for each of these conditions. The open grassy plains, sometimes verdant, at others parched up and sterile, offer an asylum to several of the true *Macropi*; the hard and stony ridges and rocky crowns of the mountains are frequented by the great Osphranters; precipitous rocks are the home of the Petrogales; the mangrove-swamps and dense humid brushes are congenial to the various *Halmaturi*; in the more spiny brigaloe-scrubs the *Onychogaleæ* form their runs, and fly before the shouting of the natives when a hunt is the order of the day; among the grassy beds which here and there clothe the districts between the open plains and the mountain-ranges—the park-like districts of the country—the *Lagorchestes* sit in their "forms," like the Hare in England; and the *Bettongiæ* and *Hypsiprymni* shroud themselves from the prying eye of man and the eagle in their dome-shaped grassy nests, which are constructed on any part of the plains, the stony ridges, and occasionally in the open glades among the brushes. The species inhabiting New Guinea (the *Dendrolagus ursinus* and *D. inustus*) resort to the trees, and, monkey-like, ascend and live among the branches. Of the Filander of the same country we know little or nothing. How wonderfully are all these forms adapted to a separate and special end and purpose—an end and a purpose which cannot be seen to advantage in any but a comparatively undisturbed country like Australia—a part of the world's surface still in maiden dress, but the charms of which will ere long be ruffled and their true character no longer seen! Those charms will not long survive the intrusion of the stockholder, the farmer, and the miner, each vying with the other to obliterate that which is so pleasing to every naturalist; and fortunate do I consider the circumstances which induced me to visit the country while so much of it remained in its primitive state.

I must revert to the Kangaroos; for it will be necessary to point out the situations affected by the various genera. In the body of the work three species of true *Macropi* are figured, and others are described, but not represented. These are all inhabitants of the southern districts of Australia and Van Diemen's Land. To say that no true *Macropus*, as the genus is now restricted, would be found in Northern Australia would be somewhat unwarrantable; at the same time, I have never seen an example from thence. The genus *Osphranter*, on the other hand, the members of which, as has been before stated, are always found in rocky situations, have their representatives in the north as well as in the south, but they are not found in Van Diemen's Land. The splendid *O. rufus* is an animal of the interior, and frequents the plains more than any other species of its genus. At present, the back settlements of New South Wales, Queensland, Victoria, and South Australia are the only countries whence I have seen specimens. The great Black Wallaroo (*O. robustus*) forms its numerous runs among the rocks, and on the summits of mountains bordering the rivers Mokai and Gwydyr. The *O. Parryi* ranges over the rocky districts of the headwaters of the Clarence and adjacent rivers, while the *O. antilopinus* is as yet only known in the Cobourg Peninsula.

The smaller *Petrogalæ* differ from all the other Kangaroos, both in the form of their feet and the structure of their brushy dangling tails. With the exception of Tasmania, these rock-lovers dwell everywhere, from north to south, and from east to west. The *P. penicillata* inhabits New South Wales; the *P. xanthopus,* South Australia; the *P. lateralis,* Western Australia; the *P. concinna* and *P. brachyotis,* the north-west coast; and the *P. inornata,* the opposite rocky shores of the east.

The true Wallabies, or *Halmaturi,* are all brush animals, and are more universally dispersed than any of the other members of the entire family. Tasmania is inhabited by two species, New South Wales by at least five, South Australia by two or three, and Western Australia by the same number; while the genus is represented on the north coast by the *H. agilis.* It will be clear, then, that the arboreal districts of the south, with their thick and impenetrable brushes, are better adapted for the members of this genus than the hotter country of the north.

The *Onychogaleæ* are, *par excellence,* the most elegantly formed and the most beautifully marked members of the whole family, and they are, moreover, as graceful in their actions as in their colouring they are pleasing to the eye. One species, the *O. fræ-nata,* inhabits the brigaloe-scrubs of the interior of New South Wales and Queensland, and probably South Australia. The *O. lunata* plays the same part, and affects very similar situations, in Western Australia; while the *O. unguifera,* as far as we yet know, is confined to the north-eastern part of the continent.

The *Lagorchestes* are a group of small hare-like Kangaroos, which dwell in every part of the interior of the southern portion of the mainland, from Swan River on the west to Queensland on the east; one species has, however, been found in the northern districts—the *L. Leichardti,* as it has been named, in honour of its discoverer, the late intrepid and unfortunate ex-plorer, Dr. Leichardt. They are the greatest leapers and the swiftest runners among small animals I have ever seen; they sleep in forms, or seats, like the Common Hare (*Lepus timidus*) of Europe, and mostly affect the open grassy ridges, particularly those that are of a stony character. The beautiful *L. fasciatus* of Swan River is one of the oldest known; the *L. Leichardti* the latest yet discovered.

The *Bettongiæ*, with their singular prehensile tails, also enjoy a wide range, the various species composing the genus being found in Tasmania, New South Wales, Southern and Western Australia, but, so far as we yet know, not in the north. For a more detailed account of the localities favoured with the presence of these animals, and the manner in which their prehensile tails are employed in carrying the grass for their nest, I must refer to the history of the respective species, and particularly to the plate of *Bettongia cuniculus*.

The *Hypsiprymni* are the least and, perhaps, the most aberrant group of this extensive family. They inhabit the southern and most humid parts of the country, and are to be found everywhere, from Tasmania to the 15th degree of latitude on the continent in one direction, and from the scrubs of Swan River and King George's Sound to the dense brushes of Moreton Bay in the other; like most other Kangaroos, they are nocturnal in their habits, grub the ground for roots, and live somewhat after the manner of the *Peramelides*, with which, however, they have no relationship.

To render my history of this group of animals the more complete, I have included in the work the three species inhabiting New Guinea : two of these belong to the genus *Dendrolagus*, and, as their name implies, dwell among the branches of trees, and rarely resort to the ground : the third forms the genus *Dorcopsis*, of which a single species only is known; it has doubtless some peculiar habits, but these must be left for a future historian to describe; at present they are unknown.

The great family of the Kangaroos, of which what I have here written must only be regarded as a slight sketch, is well worthy the study of every mammalogist. It forms by far the most conspicuous feature in the history of Australian quadrupeds; and, numerous as are the species now known, I doubt not that others will yet be discovered when the north and north-

western provinces of the country have been more diligently ex-
plored.

The third and concluding volume is devoted to the Rodents,
Seals, and Bats, and ends with the *Canis Dingo*. These are the
only Placental animals inhabiting the land of Australia, and,
contrary to what was formerly supposed, the Rodents form no
inconspicuous feature among the quadrupeds of that country.
They are very numerous in species, and almost multitudinous in
individuals. Every traveller who has visited the interior can
testify to this fact. If exploration has been his object, the
numerous runs and tracks of these little animals must have
been frequently presented to his notice,—every grassy bed
being tenanted by its own species of *Mus*, while all the
sand-hills are run over by the same or other species, interspersed with the Jerboa-like *Hapalotides*. The sluggish river-
reaches and water-holes of nearly every part, from Tasmania
through all the southern portions of the continent, have their
muddy banks traversed by the *Hydromys*, or Beaver-Rats, as
they have been very appropriately called. Even New Zealand, a
country which it was formerly supposed never had a more highly
organized indigenous creature than a bird, has its Bats; it will
not be surprising, therefore, that the sister country of Australia
should be tenanted by numerous species of these Nocturnes; not
only are they individually very plentiful, but many distinct forms
or genera are there found. The brushes which abound in fruit-
bearing fig-trees are frequented by Vampires or *Pteropi*—a form
which appears to be mainly confined to the south-eastern and
northern portions of the country, for I have not yet seen any
examples from Tasmania, or Southern or Western Australia.
The trees in this strange country which either bear fruit or
berries are very few. Even the fruit of the stately parasitic Fig
is a mere apology for that which we are accustomed to see, and
hence but few species of these great frugivorous Bats occur in
the fauna of Australia. At the same time, the paucity of species
is amply compensated by the number of individuals; these, how-

ever, are confined to the brushes which stretch along the eastern coast. In these solitary forests they teem and hang about in thousands, frequently changing their *locale* when their food becomes scarce or has been entirely cleared off. The species I more particularly allude to is the *Pteropus poliocephalus*. The Cobourg Peninsula and other parts of the north coast are also inhabited by a species which, according to Gilbert and Leichardt, is very abundant. A third and very fine one frequents Fitzroy Island, lying off the eastern coast.

The extraordinary *Molossus australis* is a native of Victoria, and is the sole species of its genus yet discovered in Australia. The *Taphozoi* appear to be rock-loving Bats, and the single species as yet discovered is from the Peninsula of Cape York. The *Scotophili*, of which there are several species, are found in all parts of the country, from Van Diemen's Land to the most northern part of the continent.

The restricted genus *Vespertilio* is more feebly represented than the last-mentioned form, since only two species are known to exist in the country; these are very generally spread over the southern coast.

Of the leaf-nosed *Rhinolophi* I have figured three species— the *R. cervinus*, which inhabits Cape York, the *R. aurantius* (a very beautiful species) from North-western Australia, and the *R. megaphyllus* from New South Wales.

The *Nyctophili*, or Long-eared Bats, are well represented, four species, at least, frequenting every part of the continent from east to west, and also the island of Tasmania.

This, I am aware, is a very imperfect *résumé* of the *Cheiroptera* inhabiting Australia; could I have rendered it more complete, I would have done so; but it must be recollected that seventh-tenths of the country are yet unexplored.

A mere glance at the globe which stands in every school-room will show how greatly the sea preponderates over the land of this planet. Like the land, the ocean is tenanted by many remarkable animals, certain groups of which exist in one hemisphere and are not found in the other; and it is not often that even the great Cetaceans occur in both. Neither do the Seals: the equatorial region separates them most completely; that is, no species is common alike to the north and the south. I do not consider that either the Australian *Cetacea* or *Phocidæ* have been well made out, and this certainly is the part of the mammalian fauna of that country of which we know the least. I have omitted the former altogether, but it will be seen that I have figured two of the latter; these constitute two genera (*Stenorhynchus* and *Arctocephalus*); they both inhabit the shores and rocky islands of the southern portion of Australia, while the Dugong (*Halicore australis*) is, as far as I am aware, a native of the east coast only.

Whether the *Canis Dingo* be really indigenous, or has at some very remote period followed man in his migrations, is a question on which naturalists are at variance. For my own part, I am inclined to the latter theory, as being the most philosophic mode of accounting for its presence there. That Man is the latest visitant to the soil of Australia there can be little doubt: the country is far too sparsely provided with fruits and other substances necessary for his existence to favour a contrary hypothesis.

In the following list of the Australian Mammals I shall refer to the volumes in which they are contained and to the plates on which they are respectively figured, and shall moreover give any additional information I may have acquired respecting them, together with an account of the new species which have been described by other writers, but which, from my not having been able to see examples, I have not figured.

Order MARSUPIATA.

Section MONOTREMATA.

Genus ORNITHORHYNCHUS, *Blumenb.*

1. Ornithorhynchus anatinus . . . . . Vol. I. Pl. 1.
Habitat. New South Wales and Tasmania. Victoria and South Australia?

Genus ECHIDNA, *Cuv.* Porcupine Ant Eaters.

2. Echidna hystrix . . . . Vol. I. Pl. 2.
Habitat. New South Wales, Victoria, the islands in Bass's Straits. Southern and Western Australia?

3. Echidna setosa, *Cuv.* . . . . Vol. I. Pl. 3.
Habitat. Van Diemen's Land.

Genus MYRMECOBIUS, *Waterh.* Brush-tailed Ant-eater.

4. Myrmecobius fasciatus, *Waterh.* . . . Vol. I. Pl. 4.
Habitat. Western Australia, and parts of South Australia.

Genus TARSIPES, *Gerv. et Verr.*

5. Tarsipes rostratus, *Gerv. et Verr.* . . . Vol. I. Pl. 5.
Habitat. Western Australia.
Mr. Waterhouse is of opinion that this animal is most nearly allied to the *Dromiciæ,* yet he has not placed it near that form in his ' History of the Mammalia.'

Genus CHŒROPUS, *Ogilby.*

6. Chœropus castanotis, *Gray* . . . . . Vol. I. Pl. 6.
Habitat. Interior of New South Wales, South and Western Australia.

Genus PERAGALEA, *Gray.*

7. Peragalea lagotis . Rabbit Rat . . . . Vol. I. Pl. 7.
Habitat. Western Australia.

Genus PERAMELES, *Geoff.* Bandicoots.

8. Perameles fasciata, *Gray* . . . . . . Vol. I. Pl. 8.

Habitat. Interior of South Australia, Victoria, and New South Wales.

9. Perameles Gunnii, *Gray* . . . . . . Vol. I. Pl. 9.
Habitat. Van Diemen's Land.

10. Perameles myosurus, *Wagn.* . . . . Vol. I. Pl. 10.
Habitat. Western Australia.

11. Perameles nasuta, *Geoff.* . . . . . Vol. I. Pl. 11.
Habitat. New South Wales.

12. Perameles macroura, *Gould.*
Perameles macroura, Gould in Proc. Zool. Soc. part x. p. 4 ;
Waterh. Nat. Hist. of Mamm. vol. i. p. 366.
Perameles macrurus, Gray, List of Spec. of Mamm. in Coll.
Brit. Mus. p. 96.

I have not figured this animal because, although twenty-one years have passed away since my description was published, I have never seen a second example ; still I have no doubt of its being a distinct species. It greatly resembles *P. obesula* and *P. nasuta,* but differs from both in its larger tail. I transcribe my original description from the ' Proceedings of the Zoological Society ' above referred to :—

" *Corpore supra nigro et flavescenti-albo penicillato, infra sordide albo, pilis rigidis obsito ; cauda pilis parvulis parce tecta, longitudine dimidio corporis æquante, supra nigra, infra fuscescenti-alba ; auris mediocribus.*

	unc.	lin.
" Longitudo ab apice rostri ad basin caudæ. .	16	3
———— *caudæ*	7	3
———— ab apice rostri ad basin auris . .	3	4
———— *tarsi digitorumque*	3	1
———— *auris*	1	2

" *Habitat.* Port Essington."

13. Perameles obesula, *Geoff.* . . . . . Vol. I. Pl. 12.
Habitat. South coasts of Australia and Tasmania generally.

E

14. Perameles Bougainvillei, *Quoy et Gaim.*

Perameles Bougainvillei, Quoy et Gaim. Zool. du Voy. de l'Uranie, p. 56. tab. 5, et Bull. des Sci. Nat. 1824, tom. i. p. 270; Waterh. Nat. Hist. of Mamm. vol. i. p. 385.

Habitat. Péron's Peninsula; in Shark Bay, Western Australia.

Having never seen a specimen of this animal, I am unable to figure it, or to say if it be a good species.

Genus PHASCOLARCTOS, *De Blainv.*

15. Phascolarctos cinereus . . . . Vol. I. Pls. 13 & 14.
Habitat. New South Wales.

Genus PHALANGISTA, *Cuv.*

16. Phalangista fuliginosa, *Ogilby* . . . . Vol. I. Pl. 15.
Habitat. Van Diemen's Land. Victoria?

In one of the letters from my son Charles, now engaged in a geological survey of Tasmania, the following passage having reference to this animal occurs:—

" I lay down, looking up at the moon and stars, thinking of home, and dreamily listening to the crackling of the fire, when a diabolical, chattering, grunting laugh overhead makes me start up, and discover that a Sooty Opossum is making an inspection of me, with comments, from the branch above; his call is responded to by others, and a kind of concert commences, which is maintained at intervals throughout the night,—the smaller or Ring-tailed Opossums performing an active part in it also, and the ' More Pork ' (*Podargus Cuvieri*) lending a little lugubrious assistance occasionally."

17. Phalangista vulpina, *Desm.* . . . . Vol. I. Pl. 16.
Phalangista melanura, Wagn., Waterh. Nat. Hist. of Mamm. vol. i. p. 288.
————— felina, Wagn., Waterh. *ib.* p. 294.

Goö-mal, aborigines of Western Australia.

Habitat. Probably every part of Australia; certainly all its southern portions.

18. Phalangista canina, *Ogilby* . . . . . Vol. I. Pl. 17.
Habitat. New South Wales.

19. Phalangista Cookii, *Desm*. . . . . Vol. I. Pl. 18.
Ngö-ra, aborigines of Perth.
Ngork, aborigines of King George's Sound.
" This species," says Mr. Gilbert, " does not confine itself to the hollows of standing or growing trees, but is often found in holes in the ground, where the entrance is covered with a stump; it is frequently hunted out of such places by the Kangaroo-dogs. It varies very much in the colour of the fur, from a very light grey to nearly a black; in one instance I caught two, from the same hole, which exhibited the extremes of these colours."
Habitat. New South Wales.

20. Phalangista viverrina, *Ogilby* . . . . Vol. I. Pl. 19.
Habitat. Van Diemen's Land and Western Australia.

21. Phalangista lanuginosa, *Gould* . . Vol. I. Pl. 20.
Habitat. New South Wales.

Genus Cuscus, *Lacép.*

22. Cuscus brevicaudatus, *Gray* . . . . Vol. I. Pl. 21.
Habitat. The Cape York district.

Genus PETAURISTA, *Desm.*

23. Petaurista Taguanoïdes, *Desm*. . . . Vol. I. Pl. 22.
Habitat. New South Wales.

Genus BELIDEUS, *Waterh.*

24. Belideus flaviventer . . . . Vol. I. Pl. 23.
Habitat. New South Wales.

25. Belideus sciureus . . . . Vol. I. Pl. 24.
Habitat. New South Wales and Victoria.

26. Belideus breviceps, *Waterh*. . . . . Vol. I. Pl. 25.
Habitat. New South Wales and Victoria.

27. Belideus notatus, *Peters* Vol. I. Pl. 26.
Habitat. Victoria.

28. Belideus Ariel, *Gould* . . . . . . Vol. I. Pl. 27.
Habitat. Cobourg Peninsula, on the north coast of Australia.

Genus ACROBATA, *Desm.*

29. Acrobata pygmæa, *Desm.* . . . . . . Vol. I. Pl. 28.
Habitat. New South Wales and Victoria.

By some oversight the name of this species has been spelt on the plate and text *Acrobates pygmæus.*

Genus DROMICIA, *Gray.*

30. Dromicia gliriformis . . . . Vol. I. Pl. 29.
Habitat. Van Diemen's Land.

31. Dromicia concinna, *Gould* . . . . . Vol. I. Pl. 30.
Dromicia Neillii, Waterh. Nat. Hist. of Mamm. vol. i. p. 315?
Habitat. Western Australia.

32. Dromicia unicolor, *Krefft.*
Dromicia unicolor, Krefft in Proc. Zool. Soc. Jan. 22, 1863.

" Fur of a uniform mouse-colour, lighter on the sides and beneath, with a blackish patch in front of the eye.

" All the hairs are slate-grey at the base, tipped with yellowish at the back and sides, and with grey beneath; longer black hairs, tipped with white, are interspersed, except on the under side of the body. Bristles black to within one-third of the tip, which is white; a few long bristly black hairs in front and behind the eye. Tail somewhat longer than the body, prehensile, thin, showing every joint; slightly enlarged at the base, and gradually tapering; covered with a mixture of light-coloured and black hairs; apical portion about $\frac{1}{4}''$ from the tip, wide beneath.

inches.

" Length from tip to tip. . . . . $6\frac{1}{4}$
Tail . . . . $3\frac{1}{4}$
Face to base of ear . . . . $\frac{7}{6}$
Ear . . . . $\frac{1}{2}$
Arm and hands . . . . $\frac{7}{8}$
Tarsi and toes . . . . $\frac{5}{6}$

" This beautiful little creature was captured near St. Leonard's North Shore, Sydney, feeding upon the blossoms of

the Banksias, and lived a few days in captivity. In its habits
it is nocturnal. The tongue of this *Dromicia* is well adapted
for sucking the honey from the blossoms of the *Banksiæ* and
Eucalypti, being furnished with a slight brush at the tip.
This species differs from the *D. concinna* of Western Australia
in being of a uniform dark colour, without the white belly,
and having the base of the tail slightly enlarged; it is about
the same size as *D. concinna*."
Habitat. New South Wales.

Genus PHASCOGALE, *Temm*.

33. Phascogale penicillata . . . . . . Vol. I. Pl. 31.
Bäl-lard, aborigines of King George's Sound.
Habitat. New South Wales, Victoria, South Australia, and
Swan River.

34. Phascogale calura, *Gould* . . . . . Vol. I. Pl. 32.
King-goor, aborigines of Williams River.
Habitat. Interior of New South Wales and the colony of
Victoria.

35. Phascogale lanigera, *Gould* . . . Vol. I. Pl. 33.
Habitat. Interior of New South Wales.

Genus ANTECHINUS, *MacLeay*.

36. Antechinus Swainsoni . . . . Vol. I. Pl. 34.
Habitat. Van Diemen's Land.

37. Antechinus leucopus, *Gray* . Vol. I. Pl. 35.
Habitat. Van Diemen's Land?

38. Antechinus ferruginifrons, *Gould* Vol. I. Pl. 36.
Habitat. New South Wales.

39. Antechinus unicolor, *Gould* Vol. I. Pl. 37.
Habitat. New South Wales.

40. Antechinus leucogaster, *Gray* . Vol. I. Pl. 38.
Habitat. Western Australia.

41. Antechinus apicalis . . . . Vol. I. Pl. 39.
Habitat. Southern and Western Australia.

Mr. George French Angas having sent me a skin of this animal from South Australia, I am enabled to state that its range extends from Western Australia to that colony.

42. Antechinus flavipes . . . . Vol. I. Pl. 40.
  Antechinus Stuartii, MacLeay in Ann. & Mag. Nat. Hist. vol. viii. p. 242; Waterh. Nat. Hist. of Mamm. vol. i. p. 416.
  Mr. Waterhouse is of opinion that the animal described as *A. Stuartii* will prove to be identical with *A. flavipes.*
  Dasyurus minimus, Geoff. Ann. du Mus. tom. iii. p. 362?; Schreb. Säugeth. suppl. tab. 152 B. e?
  Phascogale minima, Temm. Mon. de Mamm. tom. i. p. 59? ——— *affinis,* Grey, App. to Grey's Journ. of Two Exp. of Disc. in Australia, vol. ii. p. 406.
  ——— (*Antechinus*) *minima,* Waterh. Nat. Hist. of Mamm. vol. i. p. 419.
  ——— *affinis,* Waterh. *ib.* p. 421.
  See Mr. Waterhouse's remarks on the animals indicated in the last five synonyms, Nat. Hist. of Mamm. vol. i. pp. 419, 421.
  Habitat. New South Wales; and Victoria?

43. Antechinus fuliginosus, *Gould*   Vol. I. Pl. 41.
  Habitat. Western Australia.

44. Antechinus albipes . . . . Vol. I. Pl. 42.
  Habitat. Western Australia.

45. Antechinus murinus . . . . Vol. I. Pl. 43.
  Habitat. New South Wales.

46. Antechinus maculatus, *Gould* .   Vol. I. Pl. 44.
  Habitat. Queensland.

47. Antechinus minutissimus, *Gould* .   . Vol. I. Pl. 45.
  Habitat. Queensland.

Genus PODABRUS, *Gould.*

48. Podabrus macrourus, *Gould* . . . . Vol. I. Pl. 46.
  Habitat. Darling Downs in Queensland.

49. Podabrus crassicaudatus, *Gould* . . . Vol. I. Pl. 47.
Habitat. Western and Southern Australia.

Genus SARCOPHILUS, *F. Cuv.*

aireDevil 50. Sarcophilus ursinus . . . . . . Vol. I. Pl. 48.
Habitat. Van Diemen's Land.

Genus DASYURUS, *Geoff. Native Cats*

51. Dasyurus maculatus . . . . . . Vol. I. Pl. 49.
Habitat. Van Diemen's Land, New South Wales, and Victoria.

52. Dasyurus viverrinus . . . . Vol. I. Pl. 50.
Habitat. Van Diemen's Land and Victoria.

53. Dasyurus Geoffroyi, *Gould* . . . . Vol. I. Pl. 51.
Bur-jad-da, aborigines near Perth.
Bar-ra-jit, aborigines of York and Toodyay districts.
Ngoor-ja-na, aborigines of the Vasse district.
Dju-tytch, aborigines of King George's Sound.
Mr. Gilbert was informed that the stomach of this animal is frequently found to be filled with white ants.
Habitat. South portions of the Australian continent generally.

54. Dasyurus hallucatus, *Gould* . . . Vol. I. Pl. 52.
Habitat. Northern Australia.

Genus THYLACINUS, *Temm.*

ger Wolf 55. Thylacinus cynocephalus . . . Vol. I. Pls. 53 & 54.
Habitat. Van Diemen's Land.

Genus PHASCOLOMYS, *Geoff. Wombats*

56. Phascolomys Wombat, *Pér. et Les.* Vol. I. Pls. 55 & 56.
Phascolomys platyrhinus, Owen, Cat. of Osteol. Ser. in Mus. Roy. Coll. Surg. Engl. p. 334 ?
Habitat. Van Diemen's Land, and the islands in Bass's Straits.

57. Phascolomys latifrons, *Owen* . . Vol. I. Pls. 57 & 58.
Habitat. Victoria and South Australia.

58. Phascolomys lasiorhinus, *Gould* . Vol. I. Pls. 59 & 60.
Habitat. Victoria and South Australia.

59. Phascolomys niger, *Gould.*
Habitat. South Australia?

Family MACROPODIDÆ.

Genus MACROPUS, *Shaw.*

60. Macropus major, *Shaw* . . . . Vol. II. Pls. 1 & 2.
Habitat. New South Wales, Victoria, and Van Diemen's
Land.

61. Macropus ocydromus, *Gould* . . Vol. II. Pls. 3 & 4.
Speaking of this animal, Mr. Gilbert states that, " if a
female with a tolerably large one in the pouch be pursued,
she will often by a sudden jerk throw the little creature out ;
but whether this be done for her own protection, or for the
purpose of misleading the dogs, is a disputed point. I am
induced to think the former is the case, for I have observed
that the dogs pass on without noticing the young one, which
generally crouches in a tuft of grass, or hides itself among
the scrub, without attempting to run or make its escape ; if
the mother evades pursuit, she doubtless returns and picks
it up.

" Those inhabiting the forests are invariably much darker,
and, if anything, have a thicker coat than those of the plains.
The young are at first of a very light fawn-colour, but get
darker until two years old, from which age they again become
lighter, till in the old males they become very light grey. In
summer their coat becomes light and hairy, while in winter
it is of a more woolly character. It is a very common occur-
rence to find them with white marks or spots of white about

the head, more particularly a white spot on the forehead between the eyes. A very curious one came under my notice,
having the whole of the throat, cheeks, and upper part of the
head spotted with yellowish white; and albinoes have been
frequently seen by the hunters."

Habitat. Western Australia.

62. Macropus fuliginosus .   . . . Vol. II. Pl. 5.
Habitat. South Australia.

63. Macropus melanops, *Gould*.

It will be seen that I have placed this name among the
synonyms of *M. major*; but since my remarks on that species
were written, I have seen other examples so closely accordant
with the animal described by me under the above name in the
10th part of the 'Proceedings of the Zoological Society,'
that I think there is a probability it will prove to be distinct, and therefore, for the present, I restore the animal to
the rank of a species.

Habitat. Southern and Western Australia.

✓ Genus OSPHRANTER, Gould.

Generic characters.

Muffle broad and naked; *muzzle* broad and rather short;
ears moderate, rounded at the apex; fore limbs comparatively
long and stout, and the toes and claws very strong; hind
limbs short and muscular; middle toe very large; lateral toes
but little developed; two smaller inner toes, which are united
in one common integument as in other Kangaroos, terminate
in a line with the small outer toe, or nearly so; under surface
of the feet very rough, being covered with small horny
tubercles.

The above characters, especially the great expansion of the
muzzle, the comparatively small development of the lateral
toes of the hind feet, and the greater size of the middle toe,
should, in my opinion, be regarded as generic or subgeneric
rather than specific; and it was for these reasons that I pro-

posed the new sectional title of *Osphranter*. See Proceedings of Zool. Soc. part ix. p. 80.

64. Osphranter rufus, *Gould* . . . . Vol. II. Pls. 6 & 7.
Macropus (Osphranter) pictus, Gould in Proc. Zool. Soc. part xxviii. p. 373.
Habitat. New South Wales, Victoria, and South Australia.

65. Osphranter Antilopinus, *Gould* . Vol. II. Pls. 8 & 9.
Habitat. Cobourg Peninsula, Northern Australia.

66. Osphranter Isabellinus, *Gould.*
Osphranter? Isabellinus, Gould in Proc. Zool. Soc. part ix. p. 81.

General colour bright fulvous or sandy red; fur rather short, and soft to the touch; hairs uniform in tint to the base; throat and under parts of the body white, faintly tinted with yellowish in parts; fur of the belly long and very soft; the white or whitish colouring of the under parts and the uniform fulvous colouring of the upper surface and sides of the body do not blend gradually; tail similar in colour to the upper surface, but rather paler and uniform; hair of the fore feet and toes brown in front, yellowish on the sides.

The above description was taken from an imperfect skin procured at Barrow Island, on the north-west coast of Australia, and transmitted to me by Captain Stokes of H. M. S. "Beagle," and, in my opinion, pertains to a species of which no other example has yet been sent to Europe. Under this impression I have bestowed upon it the above specific appellation.

Habitat. Barrow Island, north-west coast of Australia.

67. Osphranter robustus, *Gould* . . Vol. II. Pls. 10 & 11.
Habitat. Mountain-ranges of the interior of New South Wales.

68. Osphranter? Parryi . . . . Vol. II. Pls. 12 & 13.
Habitat. Rocky mountains of the east coast of Australia from Port Stephens to Wide Bay.

Genus HALMATURUS, *F. Cuv.* ~~Kangaroo wallaby &c.~~

69. Halmaturus ruficollis . . . . Vol. II. Pls. 14 & 15.
Habitat. New South Wales.

70. Halmaturus Bennettii . . . . Vol. II. Pls. 16 & 17.
Habitat. Van Diemen's Land.

71. Halmaturus Greyi, *Gray* . . . Vol. II. Pls. 18 & 19.
Habitat. South Australia.

72. Halmaturus manicatus, *Gould* . Vol. II. Pls. 20 & 21.
Habitat. Western Australia.

73. Halmaturus Ualabatus . . . Vol. II. Pls. 22 & 23.
Habitat. New South Wales.

74. Halmaturus agilis, *Gould* . . . Vol. II. Pls. 24 & 25.
Habitat. Northern Australia.

75. Halmaturus dorsalis, *Gray* . . Vol. II. Pls. 26 & 27.
Habitat. Interior of New South Wales.

76. Halmaturus Parma, *Gould* . . . . Vol. II. Pl. 28.
Habitat. Brushes of New South Wales.

77. Halmaturus Derbianus, *Gray* . Vol. II. Pls. 29 & 30.
Thylogale Eugenii, Gray, Mag. Nat. Hist. vol. i. new ser.
1843, p. 583.
Habitat. South Australia.

78. Halmaturus Houtmanni, *Gould.*
Halmaturus Houtmanni, Gould in Proc. Zool. Soc. part xii.
p. 31.

"Of the whole of the islands forming Houtmann's Abrolhos,"
says Mr. Gilbert, "I found only two to be inhabited by this
species, viz. East and West Wallaby Islands. On both of
these they are so numerous, and have been so little disturbed,
that they will allow of a very near approach, and may in consequence be obtained in almost any number. The male
weighs, on an average, about 12 lbs.; but several old bucks
I obtained exceeded this, the heaviest weighing 15 lbs. A
mature female weighs about 8 lbs. They appear to have no

regular season for breeding, for *all* the females had young ones in the pouch, of very small size and quite naked; and none were seen or killed less than a year old, at which age their weight is about 5 lbs.

"The *Halmaturus Houtmanni* inhabits the dense scrub growing on almost every part of the two islands above mentioned; and its runs cross and recross almost every inch of them—even the sandy beaches close to the water's edge, and among the thick scrub and Mutton-bird holes; in these runs there are little sheltered spots, beneath which they lie during the heat of the midday sun, feeding for the most part during the night. On the approach of man it does not bound off at full speed as other Kangaroos do, but very leisurely takes two or three leaps, and then remaining stationary in an erect position, looks around with evident surprise, and is then easily shot. In fact, from having been so little disturbed, it will allow itself to be run down and caught. I was enabled to catch two in this way. Four or five of my men being on shore, I directed them to surround a bush into which I saw one of these Wallabies run, when the animal, seeing itself approached on all sides, became so bewildered that, instead of attempting to escape, it thrust its head into the thick scrub and allowed us to catch it by the tail.

" One I have alive has a habit of frequently crouching down like a Hare, with its tail brought forward between and before its fore feet."

Adult Male. Face dark grizzled grey, stained with rufous on the forehead; external surface of the ear and the space between the ears dark blackish grey; sides of the neck, shoulders, fore arms, flanks, and hind legs rufous, palest on the flanks; a line of obscure blackish brown passes down the back of the neck and spreads into the dark grizzled brown of the back; throat and chest buffy white; under surface of the body grey; tail grizzled grey, deepening into black on the upper side and the extremity. Fur somewhat short, coarse, and adpressed; the base bluish grey, succeeded by rufous, then white, and the extreme tip black.

Adult Female. Similar in colour to the male, but of a more uniform tint, in consequence of the rufous colouring of the shoulders and flanks being paler, and the grizzled appearance of the back not so bright.

Young. Dark grizzled grey approaching to black, particularly along the back.

	Adult Male.		Female.	
	ft.	in.	ft.	in.
Length from the nose to the tip of the tail	3	6	3	4
———— of tail........................	1	2¼	1	2
———— of tarsus and toes, including the nail	0	5¾	0	5¾
———— of arm and hand, including the nails	0	6	0	4
———— of face from the tip of the nose to the base of the ear	0	4¼	0	4
———— of ear........................	0	2¼	0	2¼

Notwithstanding Mr. Waterhouse's opinion that this animal is merely a variety of *H. Derbianus*, and what I have said in my account of that species tending to confirm his view of the subject, I have thought it best to append a copy of my original description taken from the examples sent home by Gilbert. Future research will determine whether it be identical with the *H. Derbianus* or distinct.

Habitat. Houtmann's Abrolhos, Western Australia.

79. Halmaturus Dama, *Gould.*

Halmaturus Dama, Gould in Proc. Zool. Soc. part xii. p. 32.

Dama, aborigines of Moore's River.

Mr. Gilbert states that this animal " is an inhabitant of the dense thickets of the interior, and is so exceedingly numerous that their tracks from thence to their feeding-grounds resemble well-worn footpaths. Its general habits and manners resemble those of the *Halmaturus Houtmanni.* Mr. Johnson Drummond informs me that it makes no nest, but merely squats in a clump of grass like a Hare; that it feeds in the night on the hills; and it is very difficult to procure specimens, as the places it frequents are so dense as to render shooting it almost impossible, nor can a dog even chase it. The only chance of obtaining it is by the aid of the natives, a number of whom walking or, rather, pushing their way through and beating the bush as they go abreast, and loudly shouting ' *wow, wow, wow,*' drive the *Damas* before them, when, by waiting in a clear space, you get the chance of a shot."

General colour of the fur grizzled brown, becoming of a reddish tint on the back of the neck, arms, and rump; face grey, washed with rufous on the forehead; outside of the ears and the space between them blackish grey; hinder legs light brown; tail grizzled grey; under surface of the body pale grey.

	ft.	in.
Length from the nose to the extremity of the tail ..	2	11
———— of tail.................................	1	2½
———— of tarsus and toes, including the nail	0	5¾
———— of arm and hand, including the nails	0	4¼
———— of face from the tip of the nose to the base of the ear	0	4
———— of ear.................................	0	2½

This animal is closely allied to, and of nearly the same size as *H. Thetidis,* but has much larger ears, and a much more dense and lengthened fur, the base of which is bluish grey, to which succeeds reddish brown, then silvery white, the extreme tips being black.

The above is the description of a female; the male will doubtless prove to be of larger size.

Habitat. Houtmann's Abrolhos and Western Australia.

80. Halmaturus gracilis, *Gould.*

Macropus gracilis, Gould in Proc. Zool. Soc. part xii. p. 103.

Face and all the upper surface of the body grizzled grey and dark brown, the grizzled appearance being produced by each hair being greyish white near the tip; sides of the neck and outer side of the limbs washed with reddish brown; margin of the anterior edge and the base of the posterior edge of the ear buffy white; line from the angle of the mouth dark brown; line along the side of the face, chin, and throat buffy white; under surface buffy grey; tail clothed with short grizzled hairs similar to those of the upper surface of the body, and with a line of black on the upper side at the apex for about one-third of its length; fur somewhat soft to the touch, grey at the base, then brown, to which succeeds white, the points of the hairs being black; there are also numerous long black hairs dispersed over the surface of the body; feet grizzled grey and rufous.

	ft.	in.
Length from the tip of the nose to the tip of the tail	2	6
———— of tail	1	1
———— of tarsi and toes, including the nail	0	5
———— of arm and hand, including the nails	0	$3\frac{1}{4}$
———— of the face from the tip of the nose to the base of the ear	0	$3\frac{1}{2}$
———— of the ear	0	$2\frac{1}{4}$

This is a very elegantly-formed little animal. In size it is somewhat smaller than *H. Derbianus,* and has much slighter fore arms.

Gilbert, who had a good knowledge of the Kangaroos, was always of opinion that this animal was quite distinct from every other species; and, from a careful examination of the single specimen he sent me, I entertain the same view; but I have not figured it because the example alluded to is the only one I have seen.

Habitat. The scrubs of the interior of Western Australia.

81. Halmaturus Thetidis, *F. Cuv. et Geoff.* Vol. II. Pls. 31 & 32.
Habitat. Brushes of New South Wales.

82. Halmaturus stigmaticus, *Gray* . Vol. II. Pls. 33 & 34.
Habitat. North-east coast of Australia.

83. Halmaturus Billardieri . . .   Vol. II. Pls. 35 & 36.
Habitat. Van Diemen's Land.

84. Halmaturus brachyurus . . .   Vol. II. Pls. 37 & 58.
Habitat. Western Australia.

Genus PETROGALE, *Gray.*

85. Petrogale penicillata, *Gray* . .   Vol. II. Pls. 39 & 40.
*Heteropus albogularis,*Jourd.Compt.Rend.Oct.1837,p.552,
and Ann. des Sci. Nat. Dec. 1837, tom. viii. p. 368?
Habitat. The rocky districts of New South Wales.

86. Petrogale lateralis, *Gould* . .   Vol. II. Pls. 41 & 42.
Habitat. Western Australia.

87. Petrogale xanthopus, *Gray* .   Vol. II. Pls. 43 & 44.
Habitat. South Australia.

88. Petrogale inornata, *Gould* .   Vol. II. Pls. 45 & 46.
Habitat. East coast of Australia.

89. Petrogale brachyotis, *Gould* . . . .   Vol. II. Pl. 47.
Habitat. North-western parts of Australia.

90. Petrogale concinna, *Gould* . . .   Vol. II. Pl. 48.
Habitat. North-western Australia.

Genus DENDROLAGUS, *Müll.*

91. Dendrolagus ursinus, *Müll.* .   Vol. II. Pl. 49.
Habitat. New Guinea.

92. Dendrolagus inustus, *Müll.* . . .   Vol. II. Pl. 50.
Habitat. New Guinea.

Genus DORCOPSIS, *Müll.*

93. Dorcopsis Bruni . . . . .   Vol. II. Pl. 51.
Habitat. New Guinea.

Genus ONYCHOGALEA, *Gray.* ~Small Silky haired Kangaroos~

94. Onychogalea unguifer, *Gould* .   Vol. II. Pls. 52 & 53.
Habitat. North-eastern parts of Australia.

95. Onychogalea frænata, *Gould* . . . Vol. II. Pl. 54.
Habitat. Interior of New South Wales.

96. Onychogalea lunata, *Gould* . . . . Vol. II. Pl. 55.
Habitat. Interior of Western Australia.

Genus LAGORCHESTES, *Gould. Hare Kangaroos.*

97. Lagorchestes fasciatus . . . . . . Vol. II. Pl. 56.
Habitat. Western and Southern Australia.

98. Lagorchestes Leporoïdes, *Gould* . . Vol. II. Pl. 57.
Habitat. South Australia.

99. Lagorchestes hirsutus, *Gould* Vol. II. Pl. 58.
Habitat. Western Australia.

100. Lagorchestes conspicillatus, *Gould* . Vol. II. Pl. 59.
Habitat. Barrow Island, North-western Australia.

101. Lagorchestes Leichardti . . . . . Vol. II. Pl. 60.
Habitat. The country bordering the Gulf of Carpentaria.

Mr. Blyth has described a species of this form under the name of *Lagorchestes gymnotus,* which he states is nearly allied to *L. conspicillatus,* and in all probability it is referable to one of the family figured in this work; but as the specimen is in the Museum of the Asiatic Society of Calcutta, it is impossible for me to determine this point. See "Report of Curator, Zoological Department, for May 1858," in 'Journ. Asiat. Soc. Bengal.'

Genus BETTONGIA, *Gray. Terbet Kanguroos.*

102. Bettongia penicillata, *Gray* . . . . Vol. II. Pl. 61.
Kangurus Gaimardi, Desm. Mamm. Supp. p. 542, sp. 842, 1822?
Hypsiprymnus Whitei, Quoy et Gaim. Voy. de l'Uranie, Zool. p. 62, pl. 10, 1824?
Kangurus lepturus, Quoy et Gaim. Bull. des Sci. Nat. Jan. 1824, tom. i. p. 271?

Hypsiprymnus Phillippi, Ogilb. in Proc. Zool. Soc. 1838,
p. 62?

———————— *formosus,* Ogilb. *ib.* p. 62?

———————— *minor* (Potoroo), Cuv. Règ. Anim. p. 185?

———————— *Hunteri,* Skull in Roy. Coll. of Surg. of
Eng.?

Habitat. New South Wales.

103. Bettongia Ogilbyi, *Gould* . . . . Vol. II. Pl. 62.
Wal-ya, aborigines of Perth and the mountain districts.
Woile, aborigines of King George's Sound.
Habitat. Western Australia.

104. Bettongia Cuniculus . . . . . Vol. II. Pl. 63.
Habitat. Van Diemen's Land.

105. Bettongia Grayi, *Gould* . . . . . Vol. II. Pl. 64.
Habitat. Southern and Western Australia.

106. Bettongia rufescens, *Gray* . . Vol. II. Pl. 65.
Habitat. New South Wales.

107. Bettongia campestris, *Gould* . . . Vol. II. Pl. 66.
Habitat. South Australia.

Genus HYPSIPRYMNUS, *Ill.* Kat Kangaroos

108. Hypsiprymnus murinus . . . Vol. II. Pl. 67.
Habitat. New South Wales.

109. Hypsiprymnus apicalis, *Gould* . . Vol. II. Pl. 68.
Habitat. Van Diemen's Land.

110. Hypsiprymnus Gilberti, *Gould* . Vol. II. Pl. 69.
Habitat. Western Australia.

111. Hypsiprymnus platyops, *Gould* . . Vol. II. Pl. 70.
Habitat. Western Australia.

Order RODENTIA.

Genus HAPALOTIS, *Licht.* *Long-Eared Rats.*

112. Hapalotis albipes, *Licht.* . . . . Vol. III. Pl. 1.
Habitat. New South Wales, Victoria, and South Australia.

113. Hapalotis apicalis, *Gould* . . . . Vol. III. Pl. 2.
Habitat. South Australia; and Van Diemen's Land?

114. Hapalotis hemileucura, *Gray* . . . Vol. III. Pl. 3.
Habitat. Interior of the North-eastern portions of Australia.

115. Hapalotis hirsutus, *Gould* . . . . Vol. III. Pl. 4.
Habitat. Port Essington.

116. Hapalotis penicillata, *Gould* . . . Vol. III. Pl. 5.
Habitat. Northern Australia.

117. Hapalotis conditor, *Gould* . . . . Vol. III. Pl. 6.
Habitat. Interior of New South Wales and Victoria.

118. Hapalotis murinus, *Gould* . . . . Vol. III. Pl. 7.
Habitat. Interior of New South Wales and South Australia.

119. Hapalotis longicaudata, *Gould* . . Vol. III. Pl. 8.
Habitat. Interior of Western Australia.

120. Hapalotis Mitchellii . . . . . . Vol. III. Pl. 9.
Habitat. Western and Southern Australia.

121. Hapalotis cervinus, *Gould* . . . . Vol. III. Pl. 10.
Habitat. The interior of South Australia.

I think it likely that *H. Mitchellii* may not be the *Dipus Mitchellii* of Ogilby, but that the true *H. Mitchellii* and my *H. cervinus* may be one and the same animal. If this should ultimately prove to be the case, the *H. Gouldii* of Gray will be the correct designation of the animal I have called *H. Mitchellii*, to which the terms *H. macrotis* and *H. Richardsoni* of Gray, on the specimens in the British Museum, will also probably be referable.

122. Hapalotis arboricola, *MacLeay.*

This is another of the Australian mammals of which I have not had an opportunity of examining specimens.

Two coloured sketches, accompanied by the following notes, were kindly transmitted to me by Mr. Gerard Krefft:—

" The only example of this rarity which has yet been obtained has been presented to the Australian Museum by W. S. MacLeay, Esq. It was caught at Elizabeth Bay, where it inhabits the lofty *Eucalypti,* and builds a nest among the branches, with leaves and twigs, like that of a bird."

" Fur rather harsh to the touch, and of a slate-grey next the skin,—the longer hairs, or outer coat, being mingled ochreous and black; sides greyish, with an admixture of ochreous yellow, which becomes darker towards the back, and has the black hairs much longer than on any other part; outer surface of the ears clothed with very short white hairs; throat and abdomen white; tail thinly clothed with dark-brown hairs; toes of the hind and fore feet covered with short white hairs."

Genus Mus, *Linn.*

123. Mus fuscipes, *Waterh.* . . . . . . Vol. III. Pl. 11.
Habitat. The southern portions of Australia generally.

124. Mus vellerosus, *Gray* . . . Vol. III. Pl. 12.
Habitat. South Australia.

125. Mus longipilis, *Gould* . . . . . Vol. III. Pl. 13.
Habitat. Banks of the Victoria River.

126. Mus cervinipes, *Gould* . . . . . Vol. III. Pl. 14.
Habitat. Brushes of the eastern parts of New South Wales.

127. Mus assimilis, *Gould* . . . . . Vol. III. Pl. 15.
Habitat. New South Wales, and probably Western Australia.

128. Mus manicatus, *Gould* . . . . Vol. III. Pl. 16.
Habitat. Port Essington.

129. Mus sordidus, *Gould*   Vol. III. Pl. 17.
Habitat. Darling Downs.

130. Mus lincolatus, *Gould* . . . . . Vol. III. Pl. 18.
Mus gracilicaudatus, Gould in Proc. Zool. Soc. part xiii. p. 77.
I now believe the animal I have thus named to be the
same as *M. lineolatus.*
Habitat. Darling Downs.

131. Mus Gouldi, *Waterh.* . . . . . . Vol. III. Pl. 19.
Habitat. The interior of New South Wales and Western
Australia, and probably of the intermediate countries.

132. Mus nanus, *Gould* . . . . . . Vol. III. Pl. 20.
Habitat. Interior of Western Australia.

133. Mus albocinereus, *Gould* . . . Vol. III. Pl. 21.
Habitat. Western Australia.

134. Mus Novæ-Hollandiæ, *Waterh.* . . Vol. III. Pl. 22.
Habitat. New South Wales.

135. Mus delicatulus, *Gould* . . . Vol. III. Pl. 23.
Habitat. Port Essington.

Genus HYDROMYS, *Geoff. Beaver Rats.*

136. Hydromys chrysogaster, *Geoff.* . . Vol. III. Pl. 24.
Habitat. Van Diemen's Land, New South Wales, Victoria,
and South Australia.

137. Hydromys fulvolavatus, *Gould* . . Vol. III. Pl. 25.
Habitat. The borders of the River Murray and Lake
Albert in South Australia.

138. Hydromys leucogaster, *Geoff.* . . Vol. III. Pl. 26.
Habitat. Banks of the Rivers Hunter and Clarence in
New South Wales.

139. Hydromys fuliginosus, *Gould* . . . Vol. III. Pl. 27.
Habitat. King George's Sound, and the waters near Perth
in Western Australia.

140. Hydromys Lutrilla, *MacLeay.*

I have never seen an example of the animal thus named by Mr. MacLeay, and of which two coloured sketches, one by Mr. G. French Angas, and the other by Mr. Gerard Krefft, were kindly sent to me by the latter gentleman; and without an inspection and comparison of it with the other species of *Hydromys,* it is quite impossible for me to say if it be really a species or not.

The following notes, by Mr. Krefft, accompanied the sketches :—

"The *Hydromys Lutrilla* was discovered by W. S. Mac-Leay, Esq., on the edge of the water in front of his beautiful seat, Elizabeth Bay. It is the only specimen yet seen, and Mr. MacLeay has presented it to the Australian Museum.

" Fur remarkably soft, and of a vinous or brownish grey next the skin, covered with dark brown and some sandy-coloured hairs on the flanks, and buffy hairs on the sides of the neck; throat and abdomen white; fore legs somewhat paler than the other parts of the body, with the exception of a brown patch on the upper surface of the feet; toes clothed with light-brown hairs; nails white; tarsi sepia-brown; whiskers black and white intermixed, the upper and longer hairs being the dark-coloured ones; tail about 7 inches long, five of which are covered with dark brown coarse hair without any white at the tip.

	inches.
" Length from tip to tip...................	17
—— of tail	7
—— of face to base of ear	2
—— of tarsi and toes	2 "

Habitat. New South Wales.

Family CHEIROPTERA.

Genus PTEROPUS, *Briss.*

141. Pteropus poliocephalus, *Temm.* . . Vol. III. Pl. 28.

Habitat. Brushes of New South Wales.

142. Pteropus conspicillatus, *Gould* . . Vol. III. Pl. 29.
Habitat. Fitzroy Island, off the eastern coast of Australia.

143. Pteropus funereus, *Temm.* . . . Vol. III. Pl. 30.
Habitat. The northern portions of Australia.

144. Pteropus scapulatus, *Peters.*
Pteropus scapulatus, Peters in Ann. and Mag. Nat. Hist.
3rd series, vol. ii. p. 231.

A description of this species has been published by Dr. W.
Peters of Berlin, in the number of the 'Annals and Magazine
of Natural History' for March 1863. As this description
did not appear until after these pages were in type, I have
had no opportunity of examining the specimen described,
and must therefore content myself with transcribing Dr.
Peters's remarks respecting it :—

"The present species nearly approaches *Pteropus medius*
in size, and is very easily distinguished from all other species
by two humeral spots" of ochreous-yellow, "and also by the
golden-yellow colour of the abundant woolly hair on the ven-
tral side of the wing-membranes, which appears near the
lumbar region, on the humeral membrane, and near the fore
arm almost to its end."
Habitat. Cape York, Northern Australia.

Genus Molossus, *Geoff.*
145. Molossus Australis, *Gray* . . Vol. III. Pl. 31.
Habitat. Victoria.

Genus Taphozous, *Geoff.*
146. Taphozous Australis, *Gould* . . . Vol. III. Pl. 32.
Habitat. Northern coasts of Australia.

Genus Rhinolophus, *Geoff.*
147. Rhinolophus megaphyllus, *Gray* Vol. III. Pl. 33.
Habitat. New South Wales.

148. Rhinolophus cervinus, *Gould* . . . Vol. III. Pl. 34.
Habitat. Cape York and Albany Island, Northern Australia.

149. Rhinolophus aurantius, *Gould* . . Vol. III. Pl. 35.
Habitat. Port Essington.

Genus NYCTOPHILUS, *Leach.*

150. Nyctophilus Geoffroyi, *Leach*   Vol. III. Pl. 36.
Habitat. Western Australia.

151. Nyctophilus Gouldi, *Tomes.*
Nyctophilus Geoffroyi . . . . . . Vol. III. Pl. 37.
Habitat. New South Wales.

152. Nyctophilus unicolor, *Tomes* .   Vol. III. Pl. 38.
Habitat. Van Diemen's Land.

153. Nyctophilus Timoriensis . .   Vol. III. Pl. 39.
Habitat. Western Australia.

154. Nyctophilus Australis, *Peters.*
Nyctophilus australis, Peters, in Abhandl. der Königl.
Akad. der Wissenschaften zu Berlin, 1860, p. 135 and Tab.
See a valuable paper on the genus *Nyctophilus,* by Dr.
Peters, in the above-mentioned Transactions of the Academy
of Berlin.

Genus SCOTOPHILUS, *Leach.*

155. Scotophilus Gouldi, *Gray* . . . . Vol. III. Pl. 40.
Habitat. New South Wales and Victoria ; and South Australia ?

156. Scotophilus morio, *Gray* . . . . Vol. III. Pl. 41.
Habitat. New South Wales and Victoria; and Western Australia ?

157. Scotophilus microdon, *Tomes* . . . Vol. III. Pl. 42.
Vespertilio Muelleri, Beck. Trans. Phil. Inst. Victoria, vol. iv.
part i. p. 41, with plate ?
Habitat. Van Diemen's Land ; and the south coast of Australia ?

Genus VESPERTILIO, *Linn.*

Family PHOCIDÆ, *Gray.*
Genus ARCTOCEPHALUS, *F. Cuv.*

Genus STENORHYNCHUS, *F. Cuv.*

Family CANIDÆ.
Genus CANIS, *Linn.*

Although I have omitted the Whales and Dugong, I cannot, in justice to Mr. Wm. Sheridan Wall, omit to call attention to his 'History and Description of the Skeleton of a

New Sperm-Whale lately set up in the Australian Museum; together with some account of a new genus of Sperm-Whales called *Euphysetes*,' published by W. R. Piddington, Sydney, 1851. In like manner, I cannot leave unpublished the following interesting letter respecting the Dugong, which has been forwarded to me by my brother-in-law, Charles Coxen, Esq., of Brisbane, Queensland:—

"The Dugong (*Halicore australis*, Owen) occurs in considerable numbers in Moreton Bay, but, I am led to believe, is not found further south. To the north it is plentiful in all the bays, such as Wide Bay, Port Curtis, Keppel Bay, &c., and all along the east and north coasts, in every situation suitable to its habits. In size it varies from six to nine feet in length, the latter being the size of a large "bull;" the weight also varies from 600 to 1000 lbs.; the girth at the largest part, just behind the flippers, is about six-eighths of the length; near the root of the tail it is very taper and small. The head is very peculiar: the eyes and ears are small; the nostrils small and oblique; the fleshy upper lip, which depends some three or four inches from the jaw, is peculiarly truncate in form, and covered with short stout bristles; the lower lip is globular, pendulous, and attached by a small neck to the jaw. The name given to the Dugong by the aborigines is *Young-un*. The flesh is greedily eaten and much sought for by them; and when they have been successful in procuring one or two, which occasionally happens, they gorge themselves in a most unseemly manner, and grease themselves all over with the fat and oil until they glisten in the sun like a roll of butter in the dog-days.

"The female, or ' cow,' exhibits much tenderness in the care of her offspring, and when injured utters a low, plaintive, snuffling sound, which appears to be understood by the calf.

"In the spring or calving-time they frequent the smaller bays and inlets of Moreton Bay, and are found feeding, in the more tranquil spots, on the *Alga* and other marine vegetable productions growing on the shoals near the mainland and the islands. During the winter months they are more frequently met with at sea, or outside the large bays. Their feeding-grounds vary from four to ten feet at high water.

" Harpooning is at present the only mode of procuring the Dugong. The aborigines are very expert in the use of the instrument, and the quickness of their sight renders them superior to Europeans for such service; but the loss of time, and consequent expense, owing to the unsettled habits of the natives, and at times the ruffled state of the water, have prevented its capture being entered upon as a business. A few years ago a party commenced setting nets on the shoals frequented by the Dugong, and for a time they answered the purpose; but the men engaged got careless, the nets were torn and destroyed by sharks and porpesses, and the affair fell to the ground.

" The oil, owing to its medicinal qualities, is in considerable demand, and very many persons have derived considerable benefit from its use; it is preferred to cod-liver oil, as being less disagreeable to the palate and more easily retained in the stomach. It is white and almost tasteless, and is occasionally used for frying fish. The quantity varies, according to the condition of the animal, from three to ten gallons. The meat is very good, is in flavour between beef and pork, and when salted is much like bacon.

" The head, back, sides, and tail are dark broccoli-brown; the belly and under part of the flippers light broccoli-brown, according to Werner's Nomenclature of Colours."

PROSPECTUS

OF

"THE BIRDS OF GREAT BRITAIN,"

BY

JOHN GOULD, F.R.S., ETC.

The Author has been induced to commence a work under the above title at the urgent request of numerous scientific Friends and Subscribers to his former publications, and he trusts he shall not disappoint them in their desire for a standard work on our native birds.

"The Birds of Great Britain" will be published in Imperial Folio, at the rate of two Parts a year, price Three Guineas each ; the Subscription will therefore be Six Guineas per annum. The precise number of Parts cannot be stated, but it is expected that the work will be completed in eight or nine years. With the last Part a full introduction to the subject will be published, together with Titles and every requisite to form the whole into five volumes; the first of which will comprise the Raptores or Birds of Prey ; the second and third, the Insessores or Perching Birds ; the fourth, the Rasores and Grallatores ; and the fifth, the Natatores or Swimming Birds.

It must not be supposed that this work is a continuation of, or similar to, the "Birds of Europe ;" the subjects are very differently treated as to the illustrations, the letterpress is much more voluminous, and the figures of the infantine states of most of the genera render it quite distinct.

The Author makes it a *sine quâ non* that the name of every Subscriber shall be given, and the Subscription paid on delivery, or at least annually ; and he trusts that no person will commence the Subscription without continuing it to its close.

Any additional information may be obtained by letter or by personal application to the Author at his house, 26 Charlotte Street, Bedford Square, London, W.C.

[*Turn over.*

August 1, 1862.

The Author's other Publications, all in Imperial Folio, are—

The Author will be happy to supply any of the above he may
have to those who may be desirous of completing their series of
his works; or of supplying, so far as lies in his power, the Parts
which may be required to perfect any one of them which may have
been left incomplete. He would particularly call attention to the
lately finished MONOGRAPH OF THE TROCHILIDÆ, OR HUMMING
BIRDS, one of the most beautiful and interesting of his Publications.

As no just conception of the character of these Works can be
formed from a Prospectus, the Author is willing to send a Part or
Parts for the inspection of those who may be desirous of possessing
them.

LONDON: PUBLISHED BY THE AUTHOR
AT 26 CHARLOTTE STREET, BEDFORD SQUARE, W.C.